Studies in Big Data

Volume 21

Series editor

Janusz Kacprzyk, Polish Academy of Sciences, Warsaw, Poland
e-mail: kacprzyk@ibspan.waw.pl

About this Series

The series "Studies in Big Data" (SBD) publishes new developments and advances in the various areas of Big Data- quickly and with a high quality. The intent is to cover the theory, research, development, and applications of Big Data, as embedded in the fields of engineering, computer science, physics, economics and life sciences. The books of the series refer to the analysis and understanding of large, complex, and/or distributed data sets generated from recent digital sources coming from sensors or other physical instruments as well as simulations, crowd sourcing, social networks or other internet transactions, such as emails or video click streams and other. The series contains monographs, lecture notes and edited volumes in Big Data spanning the areas of computational intelligence incl. neural networks, evolutionary computation, soft computing, fuzzy systems, as well as artificial intelligence, data mining, modern statistics and Operations research, as well as self-organizing systems. Of particular value to both the contributors and the readership are the short publication timeframe and the world-wide distribution, which enable both wide and rapid dissemination of research output.

More information about this series at http://www.springer.com/series/11970

Francesco Corea

Big Data Analytics:
A Management Perspective

 Springer

Francesco Corea
Rome
Italy

ISSN 2197-6503   ISSN 2197-6511   (electronic)
Studies in Big Data
ISBN 978-3-319-81786-6   ISBN 978-3-319-38992-9   (eBook)
DOI 10.1007/978-3-319-38992-9

Printed on acid-free paper

This Springer imprint is published by Springer Nature
The registered company is Springer International Publishing AG Switzerland

*To my parents, without whom I would
not be who I am today*

*To my brother, whom I admire and draw
inspiration from*

And to Lucia, who made me a better man,

And whose smile brings light into my days.

*Your love and support encouraged
me to chase my dreams,*

*And any small success I might reach
is as much yours as it is mine.*

Acknowledgements

The work is the result of the fruitful manifold conversations with a variety of different professionals and scientists with manifold backgrounds. So, even if it is impossible to personally thank everyone, the author is grateful to all the support and help in creating and designing this book. In particular, the work has been written while the author was an Anthemis Fellow, and with Anthemis professional and financial support. Part of the work has been done with many companies within the Anthemis ecosystem, thanks to the period the author spent directly working there. The author has also benefited from the advice and helpful comments of Gaia Fasso, Erica Young, Jacques Ludik, and Charles Radclyffe. However, all the work and any mistake remain author's own and the paper represents uniquely his view.

Contents

List of Figures

List of Tables

Chapter 1
Introduction

Abstract This introductory chapter will provide big data with a straight practical definition, bringing to light the reasons why it is such an important topic nowadays. Some of the most common applications will be listed, and a short literature review will be discussed. After this short introduction, the reader should be able to understand what big data and data science are, what has been done so far and what is the current state of art, as well as having an overview of the book.

A very well known article claims that ninety percent of the data out there have been generated through the last two years (SINTEF 2013), and it is now reaching an exponential growth rate way higher than what Moore identified for the transistors back in the Sixties. However, it is also true that even though we had several data before, yet the common perception was that managerial capabilities and performance achievements were purely driven by personal know-how or qualitative factors, rather than quantitative ones. Since then, the incremental value a single individual could bring to the business has been shrinking: everyone started having similar access to education, to use the same tools for processing insights, and finally end up with the same ideas. The profit margins for any business have thus been reducing, while the competition in any sector has increased, consistent with the notion of *red ocean* strategy (Kim and Mauborgne 2005). But then, one day someone realized that we had literally buildings of data, and that they had some value that could be used in order to discover a *blue ocean*. This is how we came across big data, and this is how everything started. Big data therefore came out as a necessity for innovation, as a potential solution and as disruptive driver able to lead every industry to the next phase of the business cycle.

Hence, this is a book about big data, data science, innovation, and new common trends in the industry. Although this is not a strategy book because every strategy worthy of its name needs to have a precise goal in mind, the work aims to provide a series of insights and best practices to help managers dealing with big data in their business contexts.

There are many ways to define what big data is, and this is probably why it still remains a really difficult concept to grasp. Today, someone describes big data as dataset bigger than a certain threshold, e.g., over a terabyte (Driscoll 2010), while others as whether it crashes conventional analytical tools like Microsoft Excel or not. More renowned works though identified big data as data that display features of *Variety*, *Velocity*, and *Volume* (Laney 2001; McAfee and Brynjolfsson 2012; IBM 2013; Marr 2015). And all of them have a kernel of truth, although they are inadequate to capture the essence of the phenomenon. The first idea appears to be incomplete since it is related to a pure technological issue, i.e., the analysis need overcomes the computational power of a single tool or machine. This would not explain however why big data came out few years ago and not back in the Nineties. The second opinion is instead too constraining, since it assumes that all the features have to be satisfied to talk about big data, and it also seems to identify the causes that originated big data (a huge amount of fast and diverse new data sources), rather than its characterization.

Hence, the definition that is going to be used for the sake of this work is different from the ones proposed so far (Dumbill 2013; De Mauro et al. 2015): data science is an innovative approach that consists of different new technologies and processes to extract worthy insights from low-value data that do not fit, for any reason, the conventional database systems (i.e., big data).

Often the words big data are misused, and intended in the same way we defined data science. Therefore, accordingly to general consensus, in the rest of the book we will use interchangeably *big data* and *data science,* but the reader should be aware of the real distinction in the terminology.

The data could be both structured (e.g., business revenues, country GDP, etc.) and unstructured (e.g., audio, video, etc.), directly streamed in memory or stored on disk. The methodologies are manifold (Manyika et al. 2011) and the applications they may have are potentially infinite (enhancing sales, reducing or managing risk, improving customer experience or the decision making process, etc.).

The scope of the new technologies is instead to reduce any latency, adopting a "schema on read"—that means storing without a prearranged schema, which is highly flexible, and the schema parsed at read time—rather than "on write". The new processes have to then move analytics to data in order to reduce data transfer costs—contrarily to what happens in the traditional ETL approach.

These new sets of techniques and vast availability of diverse data opened thus a wide spectrum of possibilities, and at the same time highlighted significant issues that ask for a certain degree of skepticism (O'Neil 2013): not everything can indeed be measured or quantified (Webber 2006), and modeling will never elide the idiosyncratic uncertainty of business (and life). It is therefore essential to develop a good "data discipline", and this is what really this book is going to be about.

In the last few years the academic literature on big data has grown extensively (Lynch 2008). It is possible to find specific applications of big data to almost any field of research (Chen et al. 2014). For example, big data applications can be found in medical-health care (Murdoch and Detsky 2013; Li et al. 2011; Miller 2012a, b); biology (Howe et al. 2008); governmental projects and public goods (Kim et al. 2014; Morabito 2015); financial markets (Corea 2015; Corea and Cervellati 2015). In other more specific examples, big data have been used for energy control (Moeng and Melhem 2010), anomaly detection (Baah et al. 2006), crime prediction (Mayer-Schönberger and Cukier 2013), and risk management (Veldhoen and De Prins 2014).

No matter what business is considered, big data are having a strong impact on every sector: Brynjolfsson et al. (2011) proved indeed that a data-driven business performs between 5 % and 6 % better than its competitors. Other authors instead focused their attention on organizational and implementation issues (Wielki 2013; Mach-Król et al. 2015). Marchand and Peppard (2013) indicated five guidelines for a successful big data strategy: (i) placing people at the heart of Big Data initiatives; (ii) highlighting information utilization to unlock value; (iii) adding behavioral scientists to the team; (iv) focusing on learning; and (v) focusing more on business problems than technological ones. Barton and Court (2012) on the other hand identified three different key features for exploiting big data potential: choosing the right data, focusing on biggest drivers of performance to optimize the business, and transforming the company's capabilities.

This book distances itself from previous literature: it analyses the current practices in the field and introduces a generalized standard framework for data strategy; it identifies key features; it finally provides suggestions of elements to be taken care of in a practitioner environment.

The structure of this work is as follows: the next section looks at data management, common myths around data science, and provides some strategic insights. Chapter 3 focuses instead on a selection of specific data challenges and scenarios, whilst Chap. 4 summarizes the features and the role of a data scientist. The final two chapters illustrate some interesting data trends, to conclude with suggestions on future directions for the research and business in the field.

References

Baah, G. K., Gray, A., & Harrold, M. J. (2006). On-line anomaly detection of deployed software: a statistical machine learning approach. In *Proceedings of the 3rd international workshop on Software quality assurance*, (pp. 70–77).

Barton, D., & Court, D. (2012). Making advanced analytics work for you. *Harvard Business Review, 90*(10), 78–83.

Brynjolfsson, E., Hitt, L. M., & Kim, H. H. (2011). *Strength in Numbers: How Does Data-Driven Decisionmaking Affect Firm Performance?* SSRN. Retrieved from http://ssrn.com/abstract=1819486.

Chen, M., Mao, S., Zhang, Y., & Leung, V.C. (2014). *Big data: related technologies, challenges and future prospects*, (p. 59). SpringerBriefs in Computer Science.

Corea, F. (2015). Why social media matters: the use of Twitter in portfolio strategies. *International Journal of Computer Applications, 128*(6), 25–30.

Corea, F., & Cervellati, E. M. (2015). The power of micro-blogging: how to use twitter for predicting the stock market. *Eurasian Journal of Economics and Finance, 3*(4), 1–6.

De Mauro, A., Greco, M., & Grimaldi, M. (2015). What is big data? A consensual definition and a review of key research topics. *AIP Conference Proceedings, 1644*, 97–104.

Driscoll, M. E. (2010). *How much data is big data?* [Msg 2]. Message posted to. Retrieved from https://www.quora.com/How-much-data-is-Big-Data.

Dumbill, E. (2013). Making sense of big data. *Big Data, 1*(1), 1–2.

Howe, A. D., Costanzo, M., Fey, P., Gojobori, T., Hannick, L., Hide, W., & Rhee, S. Y. (2008). Big data: The future of biocuration. *Nature, 455*(7209), 47–50.

IBM. (2013). *The Four V's of Big Data.* Retrieved from http://www.ibmbigdatahub.com/infograp hic/four-vs-big-data.

Kim, W. C., & Mauborgne, R. (2005). *Blue ocean strategy: How to create uncontested market space and make the competition irrelevant.* Boston, Mass: Harvard Business School Press.

Kim, G. H., Trimi, S., & Chung, J. H. (2014). Big-data applications in the government sector. *Communications of the ACM, 57*(3), 78–85.

Laney, D. (2001). 3D Data Management: Controlling Data Volume, Velocity, and Variety. META group Inc., 2001. Retrieved October 27, 2015 from http://blogs.gartner.com/doug-laney/files/2012/01/ad949-3D-Data-Management-Controlling-Data-Volume-Velocity-and-Variety.pdf.

Li, Y., Hu, X., Lin, H., & Yang, Z. (2011). A framework for semisupervised fea- ture generation and its applications in biomedical literature mining. *IEEE/ACM Transactions on Computational Biology and Bioinformatics (TCBB), 8*(2), 294–307.

Lynch, C. (2008). Big data: How do your data grow? *Nature, 455,* 28–29.

Mach-Król, M., Olszak, C. M., & Pełech-Pilichowski, T. (2015). Advances in ICT for business, industry and public sector. *Studies in computational intelligence,* (p. 200). Springer.

Manyika, J., Chui, M., Brown, B., & Bughin, J. (2011). *Big data: the next frontier for innovation, competition, and productivity.* McKinsey Report.

Marchand, D., & Peppard, J. (2013). Why IT fumbles analytics. *Harvard Business Review, 91*(1/2), 104–113.

Marr, B. (2015). *Big data: using smart big data, analytics and metrics to make better decisions and improve performance,* (p. 256). Wiley.

Mayer-Schönberger, V., & Cukier, K. (2013). *Big data: A revolution that will transform how we live, work, and think.* Eamon Dolan/Houghton Mifflin Harcourt.

McAfee, A., & Brynjolfsson, E. (2012). Big data: the management revolution. *Harvard Business Review, 90*(10), 6–60.

Miller, K. (2012a). Leveraging social media for biomedical research: how social media sites are rapidly doing unique research on large cohorts. *Biomedical Computation Review.* Retrieved October 27, 2015 from http://biomedicalcomputationreview.org/content/leveraging-social-media-biomedical-research.

Miller, K. (2012b). Big data analytics in biomedical research. *Biomedical Computation.* Retrieved October 27, 2015 form http://www.biomedicalcomputationreview.org/content/big-data-analytics-biomedical-research.

Moeng, M., & Melhem, R. (2010). Applying statistical machine learning to multicore volt-age and frequency scaling. In *Proceedings of the 7th ACM international conference on Computing frontiers,* (pp. 277–286).

Morabito, V. (2015). *Big data and analytics: strategic and organizational impacts,* (p. 183). Springer International Publishing.

Murdoch, T. B., & Detsky, A. S. (2013). The inevitable application of big data to health care. *JAMA, 309*(13), 1351–1352.

O'Neil, C. (2013). *On being a data skeptic.* Sebastopol, CA: O'Reilly Media.

SINTEF. (2013). *Big data, for better or worse: 90 % of world's data generated over last two years.* ScienceDaily, May 22, 2013.

Veldhoen, A., & De Prins, S. (2014). Applying Big data to risk management. *Avantage reply report,* (pp. 1–14).

Webber, S. (2006). Management's great addiction. In *Executive Action Report, The Conference Board,* (pp. 1–7).

Wielki, J. (2013). Implementation of the big data concept in organizations—possibilities, impediments, and challenges. In *Proceedings of the 2013 Federated Conference on Computer Science and Information Systems,* (pp. 985–989).

Chapter 2
What Data Science Means to the Business

Abstract Big data have been associated with some common misconceptions so far, and this chapter will help the reader in identify and understand those fallacies. It is going to be then shown the best data deployment approach, followed by an ideal internal data management process. A four-stages development structure will be provided, in order to assess the big data internal advancements, and a data maturity map will summarize a set of relevant metrics that should be considered for an efficient big data strategy.

Data are quickly becoming a new form of capital, a different coin, and an innovative source of value. It has been mentioned above how relevant it is to channel the power of the big data into an efficient strategy to manage and grow the business. But it is also true that big data strategies may not be valuable for all businesses, mainly because of structural features of the business/company itself. However, it is certain that a data strategy is still useful, no matter the size of your data. Hence, in order to establish a data framework for a company, there are first of all few misconceptions that need to be clarified:

(i) **More data means higher accuracy**. Not all data are good quality data, and tainting a dataset with dirty data could compromise the final products. It is similar to a blood transfusion: if a non-compatible blood type is used, the outcome can be catastrophic for the whole body. Secondly, there is always the risk of over fitting data into the model, yet not derive any further insight—"if you torture the data enough, nature will always confess" (Coase 2012). In all applications of big data, you want to avoid striving for perfection: too many variables increase the complexity of the model without necessarily increasing accuracy or efficiency. More data always implies higher costs and not necessarily higher accuracy. Costs include: higher maintenance costs, both for the physical storage and for model retention; greater difficulties in calling the shots and interpreting the results; more burdensome data collection and time-opportunity costs. Undoubtedly the data used do not have to be orthodox or used in a standard way—and this is where the real gain is locked in—and

© Springer International Publishing Switzerland 2016
F. Corea, *Big Data Analytics: A Management Perspective*,
Studies in Big Data 21, DOI 10.1007/978-3-319-38992-9_2

they may challenge the conventional wisdom, but they have to be proven and validated. In summary, smart data strategies always start from analyzing internal datasets, before integrating them with public or external sources. Do not store and process data just for data's sake, because with the amount of data being generated daily, the noise increases faster than the signal (Silver 2013). Pareto's 80/20 rule applies: the 80 % of the phenomenon could be probably explained by the 20 % of the data owned.

(ii) **If you want to do big data, you have to start big**. A good practice before investing heavily in technology and infrastructures for big data is to start with few high-value problems that validate whether big data may be of any value to your organization. Once the proof of concept demonstrates the impact of big data, the process can be scaled up.

(iii) **Data equals Objectivity**. First of all, data need to be contextualized, and their "objective" meaning changes depending on the context. Even though it may sound a bit controversial, data can be perceived as objective—when it captures facts from natural phenomena—or subjective—if it reflects pure human or social constructs. In other words, data can be *factual*, i.e., the ones that are univocally the same no matter who is looking at them, or *conventional/social*—the more abstract data, which earn its right to representativeness from the general consensus. Think about this second class of data as the notions of value, price, and so forth. It is important to bear this distinction in mind because the latter class is easier to manipulate or can be victim of a self-fulfilling prophecy. As stated earlier on, the interpretation of data is the quintessence of its value to business. Ultimately, both types of data could provide different insights to different observers due to relative problem frameworks or interpretation abilities (the so-called *framing effect*). Data science will therefore never be a proper science, because it will lack of full objectivity and full replicability, and because not every variable can be precisely quantified, but only approximated.

Let's also not forget that a wide range of behavioral biases that may invalidate the objectivity of the analysis affects people. The most common ones between both scientists and managers are: *apophenia* (distinguishing patterns where there are not), *narrative fallacy* (the need to fit patterns to series of disconnected facts), *confirmation bias* (the tendency to use only information that confirms some priors)—and his corollary according to which the search for evidences will eventually end up with evidences discovery—and *selection bias* (the propensity to use always some type of data, possibly those that are best known). A final interesting big data curse to be pointed out is nowadays getting known as the "Hathaway's effect": it looked like that where the famous actress appeared positively in the news, Warren Buffett's Berkshire Hathaway company observed an increase in his stock price. This suggests that sometime there exist correlations that are either spurious or completely meaningless and groundless.

(iv) **Your data will reveal you all the truth**. Data on its own are meaningless, if you do not pose the right questions first. Readapting what DeepThought says in *The Hitchhikers' Guide to the Galaxy* written by Adams many years ago, big data can provide the final answer to life, the universe, and everything, as soon as the right question is asked. This is where human judgment comes into: posing the right question and interpreting the results are still competence of the human brain, even if a precise quantitative question could be more efficiently replied by any machine.

The alternative approach that implements a random data discovery—the so-called *"let the data speak"* approach—is highly inefficient, resource consuming and potentially value-destructive. An intelligent data discovery process and exploratory analysis therefore is highly valuable, because "we don't know what we don't know" (Carter 2011).

The main reasons why data mining is often ineffective is that it is undertaken without any rationale, and this leads to common mistakes such as false positives, over-fitting, ignoring spurious relations, sampling biases, causation-correlation reversal, wrong variables inclusion or model selection (Doornik and Hendry 2015; Harford 2014). A particular attention has to be put on the causation-correlation problem, since observational data only take into account the second aspect. However, According to Varian (2013) the problem can be solved through experimentations.

In a similar fashion as in Doornik and Hendry (2015), it is here claimed the importance of the problem *formulation*, obtained leveraging theoretical and practical considerations and trying to spot what relationship deserves to be deepened further. The *identification* step instead tries to include all the relevant variables and effects to be accounted for, through both the (strictest) statistical methods and non-quantitative criteria, and verifies the quality and validity of available data. In the *analytical* step, all the possible models have to be dynamically and consistently tested with unbiased procedures, and the insights reached through the data interpretation have to be fed backward to improve (and maybe redesign) the problem formulation (Hendry and Doornik 2014).

Those aspects can be incorporated into a lean approach, in order to reduce the time, effort and costs associated to data collection, analysis, technological improvements and ex-post measuring. The relevance of the framework lies in avoiding the extreme opposite situations, namely collecting all or no data at all. The next figure illustrates key steps towards this lean approach to big data: first of all, business processes have to be identified, followed by the analytical framework that has to be used. These two consecutive stages have feedback loops, as well as the definition of the analytical framework and the dataset construction, which has to consider all the types of data, namely data at rest (static and inactively stored in a database), at motion (inconstantly stored in temporary memory), and in use (constantly updated and store in database). The modeling phase is crucial, and it embeds the validation as well, while the process ends with the scalability implementation and the measurement. A feedback mechanism should prevent an internal stasis, feeding the business process with the outcomes of the analysis instead of improving continuously the model without any business response (Fig. 2.1).

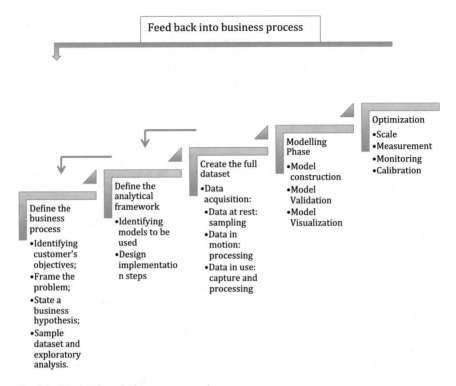

Fig. 2.1 Big data lean deployment approach

In light of all the considerations made so far, data science should then reach a compromise between the two approaches, because "*in medio stat virtus*": a specific problem should be tackled using a structured process, and an accurate question has to be posed at the beginning, but it is essential to be open and flexible to follow new unexpected paths and managing unanticipated consequences based on what data are telling. Big data are increasing the accuracy of predictions made, and enhancing the comprehension of many phenomena and human behavior. Ultimately, it appears to reduce the world complexity, providing an answer to any question posed. What is really going on instead is that they are providing multiple solutions, solutions that sometimes are so groundbreaking that they call for the question itself to be updated or amended. They disclose infinite new possibilities, which actually results in greater complexity—an intricacy that it is though manageable with a low change resiliency. Data science works as feedback-loop, and unlocking the data potential may involve a fully mind-shifting, which is important to be understood before embarking on it.

Data allow to grasp things that elude human's attention, but since they are not good or bad per se, they should never be blindly trusted. Data identification and interpretation is where the additional business value lies, and also how the mistakes or frauds occur. Value-generation is a three-steps process: it results from first determining who are the recipients of data, then correctly enquiring, and finally providing user-friendly outcomes to the right audience (and sometime great visualizations are not useful for the sake of clarity), and translating those results into actionable points.

The four misconceptions about big data summarized above seem to be the most common traps for businesses moving into this area for the first time. Extra care is suggested when big data are approached in the first place. However, even for field veterans, implementing a successful data strategy may be cumbersome, due to a set of problems experienced at different levels (technical, business, or operational) and with different degrees of intensity. On top of everything the greatest complication is the cultural issue, and how C-level professionals perceived big data projects. The top management may indeed not be aware of the potential impact of data science for their business, so they have to be instructed through a proof of concept, i.e., a short, high-value, and low-resource-consuming internal project that can persuade them on incrementing the functional area of analytics. Moreover, a second imperative is the creation of a *golden record*, which is a unique and well-defined version of every data entity, and the identification of the correct technology and architecture. This is basically a theoretical matching issue, and it has to be thought as choosing the best unambiguous key in order to understand and validate one entity and separate it from other similar ones. In this respect, it is therefore essential to establish a solid internal data procedure, which has to consist of at least four main pillars: aggregation, integration, normalization, and finally validation, as summarized in the following figure (Fig. 2.2).

Data need to be consistently aggregated from different sources of information, and integrated with other systems and platforms; common reporting standards should be created—the master copy—and any information should need to be eventually validated to assess accuracy and completeness. Finally, assessing the skills and profiles required to extract value from data, as well as to design efficient data

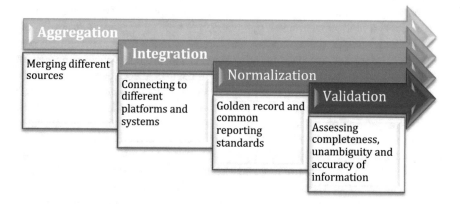

Fig. 2.2 Internal data management process

value chains and set the right processes, are two other essential aspects. Having a solid internal data management, jointly with a well-designed golden record, helps to solve the huge issue of *stratified entrance*: dysfunctional datasets resulting from different people augmenting the dataset at different moments or across different layers.

The answers to these problems are not trivial, and we need a frame to approach them. A *Data Stage of Development Structure* (DS2) is a maturity model built for this purpose, a roadmap developed to implement a revenue-generating and impactful data strategy. It can be used to assess the current situation of the company, and to understand the future steps to undertake to enhance internal big data capabilities.

Table 2.1 provides a four by four matrix where the increasing stages of evolution are indicated as *Primitive*, *Bespoke*, *Factory*, and *Scientific*, while the metrics they are considered through are *Culture*, *Data*, *Technology*, and *Talent*. The final considerations are drawn in the last row, the one that concerns the financial impact on the business of a well-set data strategy.

Stage one is about raising awareness: the realization that data science could be relevant to the company business. In this phase, there are neither any governance structures in place nor any pre-existing technology, and above all no organization-wide buy-in. Yet, tangible projects are still the result of individual's data enthusiasm being channeled into something actionable. The set of skills owned is still rudimental, and the actual use of data is quite rough. Data are used only to convey basic information to the management, so it does not really have any impact on the business. Being in this stage does not mean being inevitably unsuccessful, but it simply shows that the projects performance and output are highly variable, contingent, and not sustainable. The second Phase is the reinforcing: it is actually an exploration period. The pilot has proved big data to have a value, but new competences, technologies and infrastructures are required—and especially a new data governance, in order to also take track of possible data contagion and different actors who enter the data analytics process at different stages. Since management contribution is still very limited, the potential applications are relegated to a single department or a specific function. The methods used although more advanced than in Phase one are still highly customized and not replicable. By contrast, Phase three adopts a more standardized, optimized, and replicable process: access to the data is much broader, the tools are at the forefront, and a proper recruitment process has been set to gather talents and resources. The projects benefit from regular budget allocation, thanks to high-level commitment of the leadership team. Step four deals with the business transformation: every function is now data-driven, it is lead by agile methodologies (i.e., deliver value incrementally instead of at the end of the production cycle), and the full-support from executives is translated into a series of relevant actions. These may encompass the creation of a Centre of Excellence (i.e., a facility made by top-tier scientists, with the goal of leveraging and fostering research, training and technology development in the field), high budget and levels of freedom in choosing the scope, or optimized cutting-edge technological and architectural infrastructures, and all these bring a real impact on the revenues' flow. A particular attention has to be especially put on data lakes,

Table 2.1 Data stage of development structure

Drivers/stages	Primitive	Bespoke	Factory	Scientific
Culture	• No leadership support • Analytics as an IT asset • Conveying information (reporting, dashboard, etc.) • No budget • Descriptive analytics	• Leadership interest and midlevel management backing • Analytics used to understand problems • Specific application/department • Funding for specific project • Tailored modus operandi (not replicable) • Predictive analytics	• Leadership sponsorship • Analytics used to identify issues and develop actionable options • Alignment to the business as a whole • Specific budget for analytics function • Advanced data mining • Prescriptive analytics	• Full executive support • Data-driven business • Cross-department applications • Substantial infrastructural, human, and technology investments • Advanced data discovery • Automated analytics
Data	• Absence of a proper data infrastructure • Disorganized and dispersed silos • Duplicated information	• Data marts (lack of variety) • Internal structured data points • Data gaps or incomplete	• Virtual data marts • Internal and external data, • Mainly structured data • Easy-to-manage unstructured data (e.g., texts)	• Data lakes • Any data (unstructured, semi-structured, etc.) • Variety of sources (IoT, Social media, etc.) • Information life cycle in place
Technology	• Absence of data governance • No forefront technology (spreadsheet for reporting) • Low investments	• Integrated relational database (SQL) • Improvements in data architecture • Setting of a golden record • Scripting languages	• Pioneering technologies (Hadoop, MapReduce—see Appendix I) • Integration with programming languages • Visualization tools	• Centralized dataset • Cloud storage • Mobile applications • APIs, internet of things, and advanced machine learning tools
Talent	• Dispersed talents • Few people with few data analytical skills	• Mix of few full-time and some part-time data scientists • Proper data warehouse team • Strategic partnership for enhancing capabilities	• Well-framed recruitment process • Proper data science team • IT department fully formed and operative • Supporting of IT to data team	• Centre of excellence • Dominion experts and specialists • Training and continuous learning • Active presence within the Data Ecosystem
Impact	*No return on investments (ROI)*	*Moderate revenues, that justify though further investments*	*Significant revenues*	*Revolutionized business model (blue ocean revenues)*

repositories that store data in native formats: they are low costs storage alternatives, which supports manifold languages. Highly scalable and centralized stored, they allow the company to switch without extra costs between different platforms, as well as guarantee a lower data loss likelihood. Nevertheless, they require a metadata management that contextualizes the data, and strict policies have to be established in order to safeguard data quality, analysis, and security. Data have to be correctly stored, studied through the most suitable means, and to be breach-proof. An information life cycle has to be established and followed, and it has to take particular care of timely efficient archiving, data retention, and testing data for the production environment.

A final consideration has to be spared about cross-stage dimension "culture". Each stage has associated a different kind of analytics, as explained in Davenport (2015). Descriptive analytics concerned what happened, predictive analytics is about future scenarios (sometime augmented by diagnostic analytics, which investigates also the causes of a certain phenomenon), prescriptive analytics suggests recommendations, and finally automated analytics are the ones that take action automatically based on the analysis' results.

Some of the outcomes presented so far are summarized in Fig. 2.3. The following chart shows indeed the relation between management support for the analytics function and the complexity and skills required to excel into data-driven businesses. The horizontal axis shows the level of commitment by the management (high vs. low), while the vertical axis takes into account the feasibility of the project undertaken—where feasibility is here intended as the ratio of the project complexity and the capabilities needed to complete it. The intersection between feasibility of big data analytics and management involvement divides the matrix into four quarters, corresponding to the four types of analytics. Each circle identifies one of the four stages (numbered in sequence, from I—*Primitive*, to IV—*Scientific*). The size of each circle indicates its impact on the business (i.e., the larger the circle, the higher the ROI). Finally, the second horizontal axis keeps track of the increasing data variety used in the different stages, meaning structure, semi-structured, or unstructured data (i.e., IoT, sensors, etc.). The orange diagonal represents what kind of data are used: from closed systems of internal private networks in the bottom left quadrant, to market/public and external data in the top right corner.

Once the different possibilities and measurements have been identified—in the Appendix II, a Data Science Maturity Test (DMST) is provided, and it can be used to understand what stage a firm belongs to—it would be also useful to see how a company could transition from one level to the next. In the following figure (Fig. 2.4) some recommended procedures have been indicated to foster this transition.

In order to smoothly move from the *Primitive* stage to the *Bespoke*, it is necessary to proceed by experiments run from single individuals, who aim to create proof of concepts or pilots to answer a single small question using internal data. If these questions have a good/high value impact on the business, they could be acknowledged faster. Try to keep the monetary costs low as possible (cloud, open source, etc.), since the project will be already expensive in terms of time and

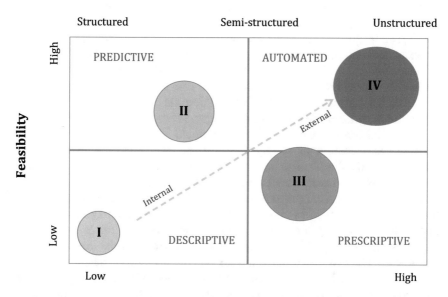

Fig. 2.3 Big data maturity map

manual effort. On a company level, the problem of data duplication should be addressed. The transition from *Bespoke* to *Factory* instead demands the creation of standard procedures and golden records, and a robust project management support. The technologies, tools, and architecture have to be experimented, and thought as they are implemented or developed to stay. The vision should be medium/long term then. It is essential to foster the engagement of the higher-senior management level. At a higher level, new sources and type of data have to be promoted, data gaps have to be addressed, and a strategy for platforms resiliency should be developed. In particular, it has to be drawn down the acceptable data loss (*Recovery Point Objective*), and the expected recovered time for disrupted units (*Recovery*

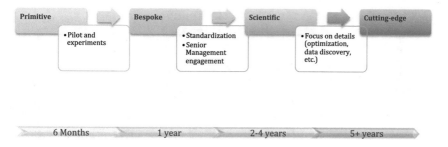

Fig. 2.4 Maturity stage transitions

Time Objective). Finally, to become data experts and leaders and shifting to the *Scientific* level, it is indispensable to focus on details, optimize models and datasets, improve the data discovery process, increase the data quality and transferability, and identify a blue ocean strategy to pursue. Data security and privacy are essential, and additional transparency on the data approach should be released to the shareholders. A Centre of Excellence (CoE) and a talent recruitment value chain play a crucial role as well, with the final goal to put the data science team in charge of driving the business. The CoE is indeed fundamental in order to mitigate the short-tem performance goals that managers have, but it has to be reintegrated at some point for the sake of scalability. It would be possible now to start documenting and keeping track of improvements and ROI. From the final step on, a process of continuous learning and forefront experimentations is required to maintain a leadership and attain respectability in the data community.

In Fig. 2.4 it has also been indicated a suggested timeline for each step, respectively up to six months for assessing the current situation, doing some research and starting a pilot; up to one year for exploiting a specific project to understand the skills gap, justify a higher budget allocations, and plan the team expansion; two to four years to verify the complete support from every function and level within the firm, and finally at least five years to achieving a fully-operationally data driven business. Of course the time needed by each company varies due to several factors, so it should be highly customizable.

Few more words should be spent regarding the organizational home (Pearson and Wegener 2013) for data analytics. We claimed that the Centre of Excellence is the cutting-edge structure to incorporate and supervise the data functions within a company. Its main task is to coordinate cross-units activities, which embeds: maintaining and upgrading the technological infrastructures; deciding what data have to be gathered and from which department; helping with the talents recruitment; planning the insights generation phase, and stating the privacy, compliance, and ethic policies. However, other forms may exist, and it is essential to know them since sometimes they may fit better into the preexisting business model (Fig. 2.5).

The figure shows different combinations of data analytics independence and business models. It ranges between business units (BUs) that are completely independent one from the other, to independent BUs that join the efforts in some

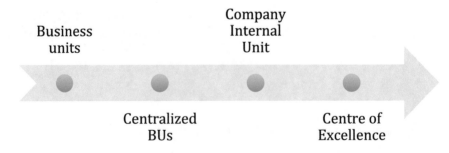

Fig. 2.5 Data analytics organizational models

specific projects, to an internal (corporate center) or external (center of excellence) center that coordinates different initiatives.

In spite of everything, all the considerations made so far mean different things and provide singular insights depending on the firm's peculiarities. In particular, the different business life cycle phase in which the company is operating deeply influences the type of strategy to be followed, and it is completely unrelated to the maturity data stage to which they belong (i.e., a few months old company could be a *Scientific* firm, while a big investment bank only a *Primitive* one).

Hence, there are at least two different approaches that have to be analyzed, i.e., the *prospective* and the *retrospective* one. The prospective concerns mainly start-ups, i.e., companies that are running on since few time and that are not producing huge amount of data (yet). They will begin producing and gathering data though, so it is extremely relevant to set an efficient data strategy form the beginning. The second case instead, the retrospective, is about existing businesses that are overwhelmed by data, and they do not know how to use them—or they may have specific problems, as centralized integration. It is clear the difference between those two circumstances, and it is then needed to explore it further.

Firstly, a startup is completely free from any predetermined structure, and it can easily establish a strong internal data policy from the beginning adopting a long-term vision, which would prevent any data related issue in the future. This is a matter to not be underestimated, and it requires an initial investment of resources and time: if the firm does it once and well, it will get rid of a lot of inconveniences and bothers. A well-set data policy would indeed guarantee a lean approach for the startup throughout any following stage. Moreover, young companies are often less regulated, both internally (i.e., internal bureaucracy is lower) and externally (i.e., compliance rules and laws). They do have a different risk appetite, which pushes them to experiment and adopt forefront technologies. Nonetheless, they have to focus on quality data rather than quantity data to start with.

A mature company instead usually faces two main issues: they have piles of data, and they do not know how to use them, or on the other side they have the data and a specific purpose in mind, but they cannot because of poor data quality, inadequate data integration, or shortage of skills. In the first case, they are in the *Primitive* stage, meaning that they have data but no clue on how extracting value from them. Since big institutions usually have really tight and demanding job roles, it is sometime impossible to innovate internally—in other words, they are "*too busy to innovate*". Some sector is more affected by this problem (fintech for instance) with respect to others (biopharma industry), as showed in Corea (2015). They should then hire a business idea generator in a first place, meaning an experienced high-level individual who can provide valuable insights even without owing a strong technical background, and afterwards a proper data scientist (or outsourcing the analysis phase). The second scenario, the one in which the data are present but useless, mainly two solutions can be adopted for each of the problems above identified: either the firm implements from scratch a new platform-team-culture, or it outsources it to intermediaries. Whereas in the first case the marginal utility has to compensate for the implementation and running costs of the new platform and the salaries for the new employees, using specialized startups, universities, or expert consultants could be quite useful due to their high specialization and flexibility. The first case is also challenging because sometime the ability to assess data is fairly poor, and the database are so disorganized and low-quality that the decision whether to invest a lot of money in something that can—but also cannot with a good probability—have a return in five years time is really terrifying.

When it comes to choose whom to outsource to, the universities often represent a preferred avenue by big corporations: universities always need funding and above all data for running their studies (and publishing their works). They cost far less than startups, they have a good poll of brains, time, and willingness to analyze messy datasets, and universities that pursue pure theoretical research can be integrated by real impactful business questions. Startups instead are revenue-generating entities, and by definition they will cost more to big incumbents, but they often gather the best minds and talents with good compensation packages and interesting applied researches that universities cannot always offer. In both cases, the biggest issue is anyway about data security, confidentiality, and privacy: what data does the company actually outsource, how the third parties keep the data secured, how do they store them, and the HiPPO (i.e., *highest paid person's opinion*) concern, are the most common issues to deal with. Another relatively new and interesting way for big corporations to get some analysis for free are meetups and hackathons, that can be exploited as window-dressing for the firm, and in the meantime used to get good analysis and insights virtually for free.

The alternative to the outsourcing scenario is the buy-in mentality, which looks at buying and integrating (horizontally or vertically) anything that the company does not develop in-house. It is definitely more costly than other options, but it solves all the problems related to data privacy and security. Incubators and accelerators can offer a substitute way to invest less in more companies of interests that deal with several useful subjects without fully buying many companies, and this is why it is becoming so popular nowadays. The disadvantage of this fragmented investment business however is that new companies have a high-risk of failing— and the big firm loses not only the amount invested but also business opportunities and competitiveness—and companies need to invest in a team to select and support the on-boarded ventures.

Hence, it could be useful to integrate all the solutions provided so far and identify a solution in the middle. What we propose here is a two-steps approach: in the first phase, universities can be used to run a pilot, or the first two to three worthy projects that can drive the business from a *Primitive* stage to a *Bespoke* one. Then, the results are used to persuade management to invest into data analytics. The perfect hybrid option would be to create an internal data analytics center that is completely both physically and administratively disconnected from the main company. Hence, in a different building, the team should be run as it was a proper startup, and has to be charged by fully autonomy and freedom of means and thinking.

The conclusions of this chapter are drawn by final note for big corporation and their data approach, as well as a list of reasons of why big data projects may fail.

It is not clear when a company should start worrying about switching or going for a big data strategy. Of course there is not a unique standard answer, because the solution is tightly related to business specificities, but broadly speaking it is necessary to start thinking about big data when every source of competitive advantage is fading away or slowing down, i.e., when the growth of revenues, clients acquisitions, etc., reaches a plateau. Big data are drivers of innovation, and this approach could actually be the keystone to regain a competitive advantage and to give new nourishment to the business. However, it should be clear by now that this is not something that may happen overnight, but it is rather a gradual cultural mind-shift that requires many small steps to be undertaken.

Concerning how and why a big data projects may fail, there could be several different reasons. There are though some more commons mistakes made by companies trying to implement data science projects. It happens often indeed that the scope is inaccurate because of lacking of proper objectives or too high ambitions. On the other hand, the excessive costs and time employed in developing efficient project result from high expectations as well as absence of scalability. Managing correctly expectations and metrics to measure the impact of big data into the business is essential to succeed in the long term.

References

Carter, P. (2011). *Big data analytics: Future architectures, Skills and roadmaps for the CIO*. IDC White Paper. Retrieved from http://www.sas.com/resources/asset/BigDataAnalytics-FutureArchitectures-Skills-RoadmapsfortheCIO.pdf.

Coase, R. H. (2012). *Essays on economics and economists*. University of Chicago Press.

Corea, F. (2015). What Finance Can Learn from Biopharma industry: an innovation models transfer. *Expert Journal of Finance, 3*, 45–53.

Davenport, T. H. (2015). The rise of automated analytics. *The Wall Street Journal*, January 14, 2015. Retrieved October 30, 2015 from http://www.tomdavenport.com/wp-content/uploads/The-Rise-of-Automated-Analytics.pdf.

Doornik, J. A., & Hendry, D. F. (2015). Statistical model selection with big data. *Cogent Economics & Finance, 3*, 1045216.

Harford, T. (2014). *Big data: Are we making a big mistake?* Financial Times. Retrieved from http://www.ft.com/cms/s/2/21a6e7d8-b479-11e3-a09a-00144feabdc0.html#ixzz2xcdlP1zZ.

Hendry, D. F., & Doornik, J. A. (2014). *Empirical model discovery and theory evaluation*. Cambridge, Mass.: MIT Press.

Pearson, T., & Wegener, R. (2013). *Big data: the organizational challenge*. Bain & Company White paper.

Silver, N. (2013). *The Signal and the Noise: The Art and Science of Prediction*. Penguin.

Varian, H. (2013). Beyond *big data*. NABE *annual meeting*. San Francisco, CA, September 10th, 2013.

Chapter 3
Key Data Challenges to Strategic Business Decisions

Abstract Many strategic challenges come with formulating a big data strategy: how to guarantee a secured data access and a constant protection of users' data; how to promise a fair data treatment; how to manage data in case of special situation such as initial public offerings, growth strategies, mergers and acquisitions; how to handle data in emerging and growing markets. Furthermore, the idea of a data ecosystem will be sketched out in this chapter.

It should be clear at this stage the degree of complexity associated to any data problem. In addition to what we showed above in terms of both common mistakes and strategic complications, specific scenarios might enhance the barriers to a correct data strategy implementation, and then have to be considered separately.

3.1 Data Security, Ethic, and Ownership

Data security and privacy represents one of the main problems of this data-yield generation, since a higher magnitude of data is correlated with a loose control and higher fraud probability, with a higher likelihood of losing own privacy, and becoming target of illicit or unethical activities. Today more than ever a universal data regulation is needed—and some steps have already been taken toward one (OECD 2013). This is especially true because everyone claims privacy leakages, but no one wants to give up on the extra services and customized products that companies are developing based on our personal data.

It is essential to protect individual privacy without erasing companies' capacity to use data for driving businesses in a heterogeneous but harmonized way. Any fragment of data has to be collected with prior explicit consent, and guaranteed and controlled against manipulation and fallacies. A privacy assessment to understand how people would be affected by data is crucial as well.

There are two important concepts to be considered from a data protection point of view: fairness and minimization. Fairness concerns how data are obtained, and the

© Springer International Publishing Switzerland 2016
F. Corea, *Big Data Analytics: A Management Perspective*,
Studies in Big Data 21, DOI 10.1007/978-3-319-38992-9_3

transparency needed from organizations that are collecting them, especially about their future potential uses. Data minimization regards instead the ability of gathering the right amount of data. Although big data is usually intended as "all data", and even though many times relevant correlations are drawn out by unexpected data merged together, this would not represent an excuse for collecting every data point or maintain the record longer than it is required to. It is hard to distinguish in this case the best practice, and until a strict regulation will not be released, an internal business practice related to industry common sense has to be used.

People may not be open to share for different reasons, either lack of trust or because they have something to hide. This may generate an adverse selection problem that is not always considered, because clients who do not want to share might be perceived as they are hiding something relevant (and think about the importance of this adverse selection problem when it comes to governmental or fiscal issues). Private does not mean necessarily secret, and shared information can still remain confidential and have value for both the parts—the clients and the companies.

This is a really complicated matter, especially if we ponder for the possibilities of a stricter regulation to raise excessive data awareness in individuals. In the moment in which customers will understand the real value laying in their personal data, they will start becoming more demanding, picky, selective, and eager to maximize their own data potential. This will probably impact some businesses and industries, which are currently based on the freeconomis model—the clients incur in zero costs in exchange of alternative sources of revenue for the firm ("*if it is free, it means you are the product*").

An important achievement regarding this matter could be reached for instance using blockchain—and an example is the use for auditing purposes. The blockchain is basically made by three components, i.e., a distributed database, an *append-only* structure, and a cryptographic secure write permissions system. This would be translated into allowing two people to have a conversation (i) without needing a server, (ii) without knowing each other and having to verify if they are who they claim to be, and finally (iii) to make the conversation public without unpleasant consequences. In other words, the blockchain technology will secure the data, it will confirm the parties' existence, and provide tailored access to specific piece of information to different actors. The important effects of the use of blockchain in relation to big data is that the data could be verified once-for-all at the beginning, and therefore is often not necessary to transfer them anymore—even within virtual spaces such as the clouds. In this way, some verified personal data will not rely anymore on the original maker, but they will become truth in stone. This would improve the configuration of the trust relation between data providers, users, and final clients, eliminating the loop structure that would consider enhancing the network trust through a control mechanism.

A special attention has to be put as well on new technologies, as for instance Hadoop. It was not designed taking into consideration a high-security level, and this is the reason why it could be highly subjected to unauthorized access and why it generated problems related to the data origin.

But this is also why organizations have to adopt a security-centric approach, and focus on increasing the security of their infrastructures (especially in distributed computing environments), protecting sensitive information and implementing a real-time monitoring and auditing process. In particular, it may be necessary to structure accesses on layers, asking for authorization and guarantying privileged access for specific users.

Bigger repositories increase the risks of cyber attacks because they come with higher payoffs for hackers or tech scoundrels. Each different source contributes to big data repositories, which means different point of access to be secured, and therefore a good infrastructure should be able to balance a flexible data extraction and analysis with a restricted unauthorized access technology. Besides, the cloud is more likely to be attacked. The server configurations may not be consistent, and gaps can be found in them, so an extra care in distributed servers is recommended.

Several solutions exist to some of these problems, and many others are being studied during these days. It is a vicious circle though, because for each problem a countermeasure can be found, but is never fully conclusive and always subject to new gaps. Current working solutions for enhancing a data security are: monitoring as much as possible the audit logs; establishing preventive measures (inactive accounts deactivation, maximum failed login attempts, stronger passwords, extra secured configurations for hardware and software etc.); using only secure tested open-source software.

On the other side, a big role will be played by the ethic behind big data, i.e., to what extent companies are going to push the data boundaries and to dig into people lives. A lot can be said on philosophical implications of big data and their relationship with human ethic (Zwitter 2014), but from a practical point of view, every company should create a data ethic internal guidelines, which should be publically displayed on the company website as well. They should develop a data stewardship, a code of professional conduct, which has to convey few main aspects: transparency (what data are used and how); simple design (simple adjustments to privacy settings if wanted); win-win scenario (make sure the customers get value from the data they provide). But above all, the golden rule—that may sound biblical—is *"don't collect or use personal data in a manner you wouldn't consider acceptable for you"*. Following these simple rules, guidelines for building a code of professional conducts have been provided in the Appendix V.

The final topic of interest is the issue around data ownership: does a customer keep the ownership of his own data, or he loses it in the moment he accepts the company's terms and conditions? There is not a straightforward answer to this question, but we can propose a solution: customers should retain the ownership on raw data initially provided; the use should be granted to the company, and cannot be call back unless specific circumstances that could negatively impact the individual occur; and finally, the company keeps the ownership on "constructed data", i.e., data obtained manipulating the original one, and that cannot be reverse-engineered to infer the raw data.

3.2 The Data Ecosystem

A very innovative concept is the idea of a data ecosystem, i.e., a common data center or storage system where different institutions share data. This sharing could indeed foster on the next level each single company, since it would allow them to have a higher base and deeper understanding of their businesses. On the other side though, every company have to maintain their strategic bottleneck in the value chain, and thus a well-thought data plan has to be put in place to not lose the competitive advantage coming from personal data. From a game theoretic point of view, this could create an equilibrium in which only useless or low-quality data are shared, so some sort of incentives have to be designed to encourage firms to share high-quality-relevant proprietary data.

This sharing system could be thought within industries and across industries. It is way easier to think about it in term of different industries with common needs, e.g., a bank who may need to score the credit profile of a client that share data with an energy provider, which has information on whether that client pays or not energy bills. Within the same industry, the matter is actually much more delicate, but still feasible, even if it could require a centralized action carried by the government.

The knowledge space so created will then be a distributed space, where different enterprise warehouses converge. This center should be managed in theory by a governmental institution or independent third-parties players, but in reality not everyone could build this space, and according to classic economic theory both data collection and data analysis should be done by the actor who can do that best.

3.3 Initial Public Offering

IPOs are special corporate situations in which a company goes public, and this requires both a large amount of data ex-ante to be organized and verified, and an extraordinary degree of further transparency after the listing. This is relevant from a data point of view then because an enormous amount of data have to be made available before the first listing day and gradually always more data will be produced around the firm's activities.

From a data perspective then, an IPO requires a higher degree of transparency and standardization, and it demands for new guidelines that make any documentation easily accessible and verifiable without unveiling competitive advantages drivers.

3.4 Growth Strategies: Acquisitions, Mergers, and Takeovers

Data analytics is key to merger and acquisition (m&a) transactions for mainly two reasons: synergies and targeting.

Sometimes it is required to a company to take care of other firms' datasets for business reasons, as for example in the case of mergers, acquisitions, or takeovers. These particular situations raise manifold relevant questions about what data strategy should be used in this case. It is important to assess the quality, depth, and variety of data the other company may have, because integration is highly costly in this scenario. Afterwards, three solutions are possible: discard all the other company's data, integrate them all within the internal data structure, or alternatively partially integrate the two big datasets. The first case assumes a perfect overlapping in terms of information available (e.g., same customers, same products, etc.), but a lower quality or depth in the information discarded. In the second case, the acquiree owns a complete different dataset that is worthy to be integrated, with the same quality and depth of the acquirer's one (e.g., Virgin Group owns business completely different from each other, but with many potential data interactions). In the final case, the two businesses are alike even if not completely equal, and they may have different data information, depth, or quality (e.g., an incomplete CRM datasets with partial information, or two businesses that sell similar products but with a different clients base).

Given the always higher relevance of embedding new datasets internally, during a m&a transaction the data architecture has to be taken into account, because sometimes the willingness of a full-integration could be vanished by out to date technologies or systems, which could considerably lower the total value of the deal. Still, the two core technologies could be compatible, but one of them could be not scalable. All these elements have to be noticed in order to price correctly the transaction.

On the other hand, big data could help in understanding what company to target, and the rationale behind it. Prediction models can show the market reactions to the deal, and allows an extensive comparison that may enlighten redundant assets or interesting new revenue-generating combinations.

3.5 Emerging Markets

The main issues in emerging markets are data scarcity and how to establish trusting relations to start collecting data. Aside from these problems, big data analytics is such a scalable business that the main distinctions between developed and emerging countries is shrinking down. The lowering cost of key infrastructure (e.g., the cloud, which provides both storage system and computational power) is allowing any kind of firm to be data-driven from day one. The emerging countries are also quickly becoming major producers of data, and as a consequence they are increasing their ability to analyze that data as well.

A different problem is also the limited access to relevant information and analytics that could to gauge supply and demand for certain businesses. Even getting the raw data is hard, and a lot of effort has to be done to convert unstructured data into something workable. By contrast with respect to developed countries, the majority of the digital data in emerging economies are mobile data, but it is still a small percentage compared to the non-digital data produced. Further issues may arise from intrinsic sample biases, because individuals who source the data belong to specific clusters of the population. The limited amount of talents and the scarcity of infrastructures represent another challenge.

Emerging economics are the best scenario for data-trials, but they need redesigned data strategies: the pace of development of data strategies will be slower, and it requires starting from small data first.

References

OECD. (2013). *OECD guidelines on the protection of privacy and transborder flows of personal data*, (pp. 1–27).
Zwitter, A. (2014). Big data ethics. *Big Data & Society, 1*(2), 1–6.

Chapter 4
A Chimera Called Data Scientist: Why They Don't Exist (But They Will in the Future)

Abstract This chapter discusses the role of the data scientist, what a data scientist is, and the set of skills needed to become one. The underlying idea proposed is that the job market is not matured enough yet for this figure to be trained and employed, but it will be ready in the next few years. In order to help firms to understand what to look for and how to use resources in the best way, a personality test has been implemented and different types of data scientists have been classified using this test.

All this confusion and vagueness around definitions and concepts, and the hurdles technicalities of the big data black box have turned the people who analyze huge datasets into some kind of mythological figures. These people, who possess all the skills and the willingness to crunch numbers and providing insights based on them, are usually called data scientists. They have inherited their faith in numbers from the Pythagoreans before them, so it may be appropriate to name them *Datagoreans*. Their school of thinking, the Datagoreanism, encourages them to pursue the truth through data, and to exploit blending and fruitful interactions of different fields and approaches for postulating new theories and identifying hidden connections.

However, the general consensus about who they are and what they are supposed to do (and internally deliver) is quite loose. Just by browsing job offers for data scientists one understands that employers do not often really know what they are exactly looking for, and this is probably why everyone is complaining about the shortage of data scientists in the job market nowadays (Davenport and Patil 2012).

In reality data scientists as imagined by most do not exist because it is a complete new figure, especially for the initial degrees of seniority. However, the proliferation of boot camps and structured university programs on one hand, and the companies' increased awareness about this field on the other hand, will drive the job market towards its demand-supply equilibrium: firms will understand what they actually need in term of skills, and talents will be eventually able to provide those (verified) required abilities.

It is then necessary at the moment to outline this new role, which is still half scientist half designer, and it includes a series of different skillsets and

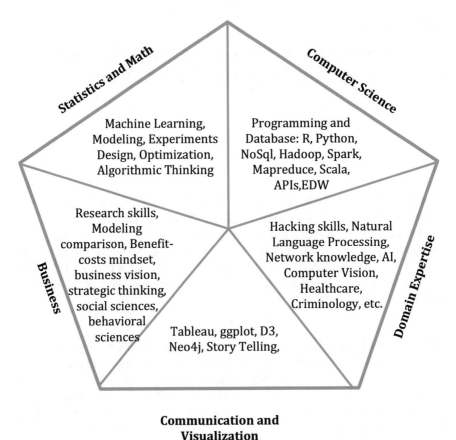

Statistics and Math

Machine Learning, Modeling, Experiments Design, Optimization, Algorithmic Thinking

Computer Science

Programming and Database: R, Python, NoSql, Hadoop, Spark, Mapreduce, Scala, APIs,EDW

Business

Research skills, Modeling comparison, Benefit-costs mindset, business vision, strategic thinking, social sciences, behavioral sciences

Domain Expertise

Hacking skills, Natural Language Processing, Network knowledge, AI, Computer Vision, Healthcare, Criminology, etc.

Tableau, ggplot, D3, Neo4j, Story Telling,

Communication and Visualization

Fig. 4.1 Data scientist core skills set

capabilities, akin to the mythological chimera. The profiling is then provided in the following table, and it merges basically five different job roles into one as shown in Fig. 4.1: the computer scientist, the businessman, the statistician, the communicator, and the domain expert (a more complete list of skills could be found in the Appendix III).

Clearly, it is very cumbersome if not impossible to substitute five different people with a single one. This consideration allows us to draw several conclusions. First, collapsing five job functions into one is efficient on one hand because the value chain is concentrated and not dispersed, but on the other hand it can demand more time and resources. Indeed, a single individual can probably be less productive than five different people working on the same problem in the same timeframe. Secondly, hiring one specialist should cost less than a total of five semi-specialists, but much more than anyone of them singularly considered because of his specialization and high-level knowledge and flexibility. Looking at some number though, this does not seem to be reflected in the current job market:

using Glassdoor.com, it is possible to notice that on average in 2015 in the United States (i) a computer scientist annually earns around $110,000, (ii) a statistician around $75,000, (iii) a business analyst $65,000, (iv) a communication manager $80,000, and finally, (v) a domain expert about $57,000. On the contrary, a data scientist salary median is around $100,000 according to the most recent survey run by O'Reilly (King and Magoulas 2015). From the survey it is possible to also learn that an average working week lasts often 40 hours, and that their daily tasks mainly consist of four hours per day spent on ETL, data cleaning, and machine learning (belonging to the computer science realm), with only a couple of hours spent on proper data analysis. According to these statistics, and roughly (and maybe incorrectly from a practitioner's point of view) assuming that the rest of their time is equally divided into the other three activities, a data scientist should earn around $92,000. This is of course a very approximate estimate, which does not into account any seniority, differences across industries, etc., and where the domain expertise is computed as the average of marketing ($55,000), finance ($65,000), database ($57,000), network ($64,000), and social media ($41,000) specializations. But it does convey a broad concept: data scientists seem to be (almost) fairly compensated in absolute terms, but their remuneration is definitely lower if compared to the cost structure they are facing to become such specialized figure. Indeed, the learning cost is really high because there are not so many designed programs for data science, so they have to do great efforts in filling their knowledge gaps. Furthermore, the universities and training monetary costs are really burdensome, and the opportunity costs quite heavy, and since the path is definitely new and not well established the choice of becoming a data scientist is risky and expensive.

All the considerations drawn so far point to a few suggestions for hiring data scientists: first of all, data science is teamwork, not a solo sport. It is important to hire different figures as part of the team, rather than exclusively for individual abilities. Moreover, if a data science team is a company priority, the data scientists *have to be hired to stay*, because managing big data is a marathon rather than 100 m dash.

Secondly, data scientists come with two different DNAs: the scientific and the creative one. For this reason, they should be let free to learn and continuously study from one hand (the science side), and to create, experiment, and fail from the other (the creative side). They will never grow systematically and at a fixed pace, but they will do that organically based on their inclinations and multi-faceted nature. It is recommended to leave them with some spare time to follow their inspirations (some company is already doing that since years, and it consists in leaving them 10–20 % of their working hours to pursue their personal ideas, to innovate, or simply for self-discovering). Furthermore, they have to be highly motivated, because often money even relevant is not anyway the crucial aspect to them. A high salary is indeed a signal of the respect the company has for their work—both past and future—and it is of course an incentive because they could potentially go working anywhere else. Although, the retention power of a good salary is quite low with respect to interesting daily challenges, and in order to align the data experts' interests with the company vision they have to be continuously fed with stimulating problems, and their work has to be relevant and impactful. Remember also that the scientist part requires them to be part of a bigger community, as well as the freedom to share concepts, ideas, and eventually working in parallel with peers. Even though the companies believe in patenting and scarce divulgation of what they do to maintain a sustainable competitive advantage, they have to compromise with the fact that scientists need to publish their researches, sharing data, materials, and ideas.

Finally, do not be closed-minded and suppress any prejudice. Even if on percentage there is a good share of American male with a PhD working in data science (King and Magoulas 2015), this may be indicative but not conclusive on the ideal candidate to hire: value the skills and capabilities more than titles or formal structured education—at least as soon as the field would be deeply-rooted and university degrees would be a good signal for skills owned. So far, in order to become a data scientist, the paths to be followed could be unconventional and various, so it is important to assess the abilities instead of deciding based on the type of background or degree level. Never forget that one of the real extra-value added by data science is different field contaminations and cross-sectional applications. It is also essential to take into account that not all the data specialists are the same (Liberatore and Luo 2012; Kandel et al. 2012), and it is possible to cluster them in four different groups (Harris et al. 2013) and by four different personalities in order to reach a higher type of granularity, based on their actual role within the company ("Archetype") and on personal features ("Personality"—according to the *Keirsey Temperament Sorter*). Identifying correctly the personality type of a data scientist is crucial to amplify his internal contribution and efficiency, as well as to maximize the resources employed to recruit him (Table 4.1).

Table 4.1 Data scientists' personality assessment and classification

Archetype/Personality	Artisan	Rational	Guardian	Idealist
Technical	*Gardener*: Data munging and coding	*Wrangler*: Algorithms implementation	*Architect*: System architecture and infrastructure	*Evangelist*: Enhancing technical community
Researcher	*Alchemist*: Experiments, exploration and ideas generation	*Groundbreaker*: Innovative methodologies and modeling	*Cruncher*: Analytical model optimization	*Champion*: Mentoring and teaching
Creative	*Trailblazer*: Spotting out new hidden connections	*Warlock*: Using new tools for new applications	*Catalyst*: Customer intelligence	*Visionary*: Information diffusion to public
Strategist	*Babelian*: Data Interpreter	*Fisherman*: Blue ocean strategy and monetization	*Mastermind*: Project management	*Advocate*: Promoting to management

In the table above, a full disentanglement of data scientists' types is provided. The color roughly represents the partition between three main skills they possess—based on the survey run by Harris et al. (2013)—that are mathematics-statistic-modeling skills (blue), business ones (green), and coding abilities (red). Having this clear classification in mind may be argued to be a merely speculative and useless labeling exercise, but it is indeed extremely relevant because increases the data science team efficiency: identifying personal inclinations and aspirations would allocate the best people to the best job role, and common complaints and problems such the insufficient time for doing analysis, the poor data quality, and the excessive time spent in collecting and cleaning data (King and Magoulas 2015) would be eliminated—or better, they would be assigned to the right people. Furthermore, this framework would help identifying the minimum team structure to start with: on the main diagonal there are indeed the basic figures needed in order to properly establish a fully-functional data science team. The *Gardener* (usually known also as data engineer) is in charge of maintaining the architecture and making the data available to the *Groundbreakers*, who are usually identified as proper data scientists, and that try to answer research questions and draw insights from data once they verified through tests that their models work. The insights are then passed to *Advocates* (business intelligence) and *Catalysts* (customer intelligence team), who respectively communicate that information to executives and use that to increase customers' satisfaction. The data process is illustrated in the following figure (Fig. 4.2).

Having these four different basic teams guarantees an efficient data-driven business and a sharp outcomes delivering, as soon as the communication across-teams and across-departments is well implemented. It is common practice to have short (five-minutes at most) stand-up internal meetings every morning to wrap up the daily objectives, works, and expected outcomes, as well as weekly meetings with other departments to align the work. Structuring the process as above proposed would finally increase the scalability of any data project.

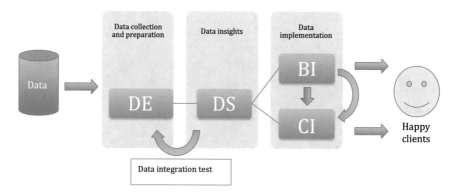

Fig. 4.2 Data science value chain

References

Davenport, T. H., & Patil, D. J. (2012). Data scientist: The sexiest job of the 21st century. *Harvard Business Review, 90*(10), 70–76.

Harris, H. D., Murphy, S. P., & Vaisman, M. (2013). *Analyzing the Analyzers*. O'Reilly Publishing.

Kandel, S., Paepcke, A., Hellerstein, J. M., & Heer, J. (2012). Enterprise Data analysis and visualization: An interview study. In *Proeedings of IEEE Visual Analytics Science & Technology (VAST)*.

King, J., & Magoulas, R. (2015). *2015 data science salary survey*. United States: O'Reilly Media, Inc.

Liberatore, M., & Luo, W. (2012). ASP, the art and science of practice: A comparison of technical and soft skill requirements for analytics and or professionals. In *Interfaces 201343*, (vol. 2, pp. 194–197).

Chapter 5
Future Data Trends

Abstract Big data and data science pushed the technological frontier one step forward, and as they are an innovation themselves, they also entail the development of new trends. The Internet of Things is the first trend highly interconnected with big data that will be discussed. Secondly, an overview of cloud technologies will be provided. Finally, application-programming interfaces will be shown to have a huge impact on how data are accessed, protected, used, and widespread.

Data are permeating our world, and they will play a decisive role in the next future as well. They are changing our way of thinking, our business models, and how we approach any disciplines. We are already observing some manifestation of these changes, and we already have evidences of important data trends, such as the data becoming a completely new asset class or a new currency as well, which may supplement one day the traditional ones. There are although other more specific trends, namely the cloud technology, APIs, and the Internet of things (IoT) that need to be tackled separately because of their impact of our daily activities. Although each one of these tendencies has been introduced for a different purpose, it is turning out that they are all driving data science toward two major achievements: they are indeed increasing the interconnectivity around us, and they are increasing our personal utility. The latter improvement is reached by reducing the frictions (e.g., cloud, APIs, etc.), enhancing more than proportionally the cost-benefit tradeoff, or making us aware of an utility increase—some time the relation action-utility is not consciously understood, and we only need to become aware of it.

Every single trend would need a complete book to be fully understood and integrated in a big data strategy, but the purpose of the next paragraphs is rather to acknowledge their existence and comprehend how they relate to the data science world.

5.1 The Internet of Things (IoT)

This is one of the reasons why nowadays big data is such a buzzword. The IoT is indeed generating and gathering data that were inaccessible few years ago. Mobile phones enlarge the spectrum of the data collectors and improve the micro-data quality tremendously; sensors are wearable by anyone and available for a really low-cost; drones and nanosatellites allow us to reach horizons that we would not be able to achieve otherwise; smart buildings and cities scale up big data to have an exponential impact, while roboadvisory reflects the frontier of new human interactions. In spite of the specific application we are dealing with, every IoT node has in common three main features: they are remote, automatic, and dynamic. Metaphorically speaking, it works like a hive with a central brain and several worker bees.

On the other side though, they are increasing the complexity geometrically, and they are raising a series of challenging situations. First of all, the Internet of Things is posing a more careful emphasis on marginal features over core aspects. Secondly, reliability assumes an always greater importance: the fostered interconnectivity augments indeed the risk of an *total collapse point*, i.e., a situation in which the presence of multiple nodes shuts down the entire system at once. Moreover, even if it may sound counterintuitive, the way in which we are structuring the nodes and single information vehicles is not through modules and neither it is fully integrated by default, as it was in the past.

Big data and IoT are closely related, both at individual and industrial level. IoT is drastically changing several business models, since it is integrating with big data techniques that collect information in real-time, using them for diversified revenue streams.

5.2 The Cloud

The cloud is rapidly becoming the company's best friend: it allows to cut down the storage costs to virtually zero and scale up the analysis, because data will be store in this ether: there is not any capital expenditure on infrastructure or maintenance, and even personnel (especially the highly specialized and thus expensive) is lowered down, and this is what often makes data science cost-effective. Reducing the concept to the ground, using a cloud means renting out a server space not physically located in your own building. In addition to the storage space, the cloud is essential in order to exploit the CPU power of different machines for tasks that a single one is not able to implement. Although is a powerful tool and represents one of the new data analytics frontiers, it is neither a totally new idea nor a bullet-proof instrument. First of all, as already pointed out for the IoT trend, it demands a careful reflection on systemic risk and fragility. Secondly, it might generate issues for people to reaccess the data, and encourages accumulating things (and data) we do not really need, but that we are not able to discard. So, one of the problem that it creates is the decision of what kind of data has to be stored. Would you store useless data, or something you are not going to use again on the cloud, or rather using it as a backup option? In the first case, why are you then storing those data if they are useless? And in the second scenario, there are not any other (maybe more efficient) options? An alternative could be, for instance, a peer-to-peer system. From an enterprise point of view instead, every company is increasing exponentially their hardware capacities and disk spaces, so why they should go for an extra virtual space when they already have plenty of unused space? Some big corporations would argue that they may use the cloud in order to create and maintain a golden record, but does it really make sense given the increasing ability of our algorithm to recognize the data entities with unsupervised methods?

Hence, when it comes to the cloud, many aspects have to be taken into account, and there are things to be considered that entirely depend from others. An example is the elasticity provided by this system—that is usually ruled by a "pay as you go" option plan—that is balanced by the fact of sharing the same server space with others (who sometimes can be "noisy neighbors", i.e., players that affect negatively others with their behavior).

5.3   Application Programming Interfaces (APIs)

Although it might not be the first objective for a successful data strategy, APIs management is essential in order to reach efficiency and scalability. The Application Programming Interfaces (APIs) represent standardized access to proprietary data. They are becoming extremely relevant to the field for a series of reasons: first of all, they lower significantly the cost and effort of accessing the data for whatever purpose they might be needed (cleansing, analysis, etc.). These lower barriers to entry are applicable also to non-technical audience and users, and this is probably the greatest contribution APIs bring. In addition, they foster and facilitate the integration of different datasets, reducing the likelihood of internal bugs within the servers. They smooth the data flow and lifecycle, providing a more scalable, reliable, and efficient connection between data, applications, devices, and eventually people.

The business and economic validity of these gateways is in the new use cases they create, and in how multi-stages APIs really create new (meta)data, unlocking data not accessible before.

Furthermore, they add a further layer of security to the data access, since they allow only to a selection of approved users to navigate through information that should not be available to everyone—think about potential application in protecting customers' rights and data, or investment banking internal firewall, or again simply different departments of a company that should not have access to the same data for legal or ethical reasons.

As downside though, a common fear is that they might weaken in the long term the human ability to perform the tasks assigned to the algorithm: the compromise is creating a system that learns from and teaches to human users at the same time.

Chapter 6
Where Are We Going? The Path Toward an Artificial Intelligence

Abstract Big data represented a topic of high interest in the last five years, and brought an innovation and technological breakthrough all around. However, they are only one step toward a more revolutionizing advancement, i.e., the Artificial Intelligence. The ability to handle a huge amount of data and the capacity to harness their power will eventually allow us to design algorithms and hardware that could at least approximate, if not fully reproduce, the human brain capabilities.

While trends must have some evidences in the present to be acknowledged, future horizons concern things possibly reachable in the future, but not today. So far, big data has been used to understand the complexity that surrounds both social and natural events. However, we are increasingly trying to emulate the human brain, as we strive for new knowledge. This project is the most complex and successful big data project ever realized: a machine that processes a huge amount of both structured and unstructured data in short time frames and act upon the results achieved.

In other words data science only represents the infancy stage of artificial intelligence (AI), i.e., the field that studies how to create a computer (or more generally machine) ables to prove an intelligent behavior. Big data is instead only the fuel that feeds AI. There are different approaches and kinds of AI, and each of them provides different insights. An AI algorithm could indeed be specific, generic, or social. The first category solves specific issues learning throughout the journey; the second one is more strictly related to the human brain functioning, which is the reason why is usually known as strong AI; finally, the social AI simulates social interactions and human affects. Hence, machine should not only be able to learn, plan, reason, deduce, and perceive through sensors, but also to feel empathy, to have intuitions, and to prove creativity.

However, the creation of a thinking machine raises several feasibility doubts and ethical issues, and it entails a series of interesting open questions. One of them is whether AI can help humans in formulating "big questions". In other words, is it possible to imagine and build a final and omnicomprehensive algorithm able to

© Springer International Publishing Switzerland 2016
F. Corea, *Big Data Analytics: A Management Perspective*,
Studies in Big Data 21, DOI 10.1007/978-3-319-38992-9_6

run all the possible analysis and correlations in order to automatically detect hidden connections and trends and suggesting what investigating further? This would exclude any kind of human contribution from the equation, but it creates another dilemma: if it is true that emotions, heuristics, personal behavior, instincts and priors sometime support the human being in fastening his decision-making process—cutting the costs and making it more efficient-, shall we then try to postulate the existence of an intelligent machine with some degree of human judgment?

These are only a few of the next level questions we should be thinking of, and the answers are not easy to find. So far, it seems that machine and human learning are complementary and compatible, and none of the two is perfectly efficient without the other one. Indeed, the real innovators in the last five years have been those ones who were able to harmonize the computational ability of the machine with the capacity of the human beings of identifying hidden patterns and drawing insights—to use Eric Brynjolfsson's words, the ones who rose with machines rather than against them.

Chapter 7
Conclusions

Big data is a solution, but is not *the* solution. It has to be wisely understood and managed, and it is not a panacea for the ills, but it can lead any business to the next level. It is a fast-pace environment, and things that work today may not apply tomorrow, so instead on concluding with strong statements written in stone, it is preferable to provide some final thoughts in form of pieces of advice.

First of all, don't think too much around big data, but rather start practicing it. They represent a new field of exploration for everyone, and a great competitive advantage to not be missed. Hence, even if it will be a low profile or low-increasing revenues project, start immediately—with small inexpensive pilots and taking care of your "small" data first—and fail, fail, fail. Fail faster, fail better; experiment and fail directly in the open market, because this is where the innovative pioneering ideas and feedbacks come from. Embrace failure as never before, because it teaches something the nothing else can convey, and build your "reputation of failure"—companies will be judged in a future on how they fail and react to failure rather that how they succeed. Create your own solutions because big data applications are not fully transferrable, so be unique in what and how you use big data, and scale up quickly.

Data are definitely a strategic priority and necessity for every firm, but the normal course of the business entails a tradeoff between surviving—doing what your daily job is and what you are good at—and innovating (and risking), but you have to be brave and adopt data science as the new company lifestyle.

Secondly, develop a data sharing discipline and a (non) ownership culture. The next business level is represented by professional ecosystems, ruled by a data democracy, so employ some effort in advancing and cultivating also open-source solutions. Create a data manifest and foster a strong internal communication, which is fundamental for big corporations, since often people do not even know who may have what, or whether some colleagues is already working on a pilot, on a specific project, or using some particular data or model.

Big data will lead to think that everything is possible and at fingertips, and maybe this will be the case in some years. Hence, (i) think holistically and without boundaries, and be a data visionary; (ii) think strategically, state a data governance

© Springer International Publishing Switzerland 2016
F. Corea, *Big Data Analytics: A Management Perspective*,
Studies in Big Data 21, DOI 10.1007/978-3-319-38992-9_7

policy, but do not let you be bounded by either too strict culture, discipline, or formal structures. Be flexible and ready to readapt quickly to what the data suggest you to go for; (iii) think complementary, since big data technologies will pair rather than replace completely your current systems; and (iv) focus on execution, prioritize your activities and set an underlying roadmap to follow. Understand where investing the capitals, what are the perceived and the actual ROI of big data applications, and whether they are enlarging the spectrum of your possibilities, lowering the costs, or reducing the problems.

But above all, embrace the continuous change and adopt flexibility as a mantra: in the big data world, *"things keep going working until they do not"*, and changing at the same pace of data analytics is the quintessence to succeed in a business context.

Appendices

Appendix I—Nomenclature for Managers

Relational database management system (RDBMS): structured data in predetermined schema (*tables*), scalable vertically through large SMP servers, or horizontally through clustering softwares. These databases are usually easy to create, access, and extend. The standard language for relational database interoperability is the **Structured Query Language (SQL)**.

Non-relational database: database that does not store data into tables, but made them accessible through special query APIs. The standard language used is **Not Only SQL (NoSQL)**: it does not present a fixed schema, it uses BASE system to scale vertically (basically available, soft-state, eventually consistent), and sharding (horizontal partitioning) to scale horizontally. Examples are **MongoDB** and **CouchDB** (they differ mainly because in MongoDB the main objects are *documents*, while in CouchDB are *collections*, which in turn contain documents). NoSQL commonly used **JavaScript Object Notation (JSON)** data format (**BSON** in MongoDB—binary JSON), and it mainly works through **Key Value Store (KSV)**, i.e., a collection of different unknown data types (while a RDBMS stores data into table knowing exactly the data type).

Hadoop: open source software for analyzing huge amount of data on a distributed system. His primary storage is called Hadoop distributed file system (**HDFS**), which duplicates the data and allocates them in different nodes. It has been written in Java. It is a core technology in the big data revolution and stores data into their native raw format, and it can be used for several purposes (Dull 2014), such as a simply data staging or landing platform complementary to the existing EDW (as an enterprise data hub, i.e., EDH), or managing data (even small), transforming those into a specific format in the HDFS and sending them back to the EDW, lowering thus the costs while increasing the processing power. Furthermore, it can integrate external data-sources and archive data (both on-premises or into the cloud), and reduce the burden for a standard EDW.

© Springer International Publishing Switzerland 2016
F. Corea, *Big Data Analytics: A Management Perspective*,
Studies in Big Data 21, DOI 10.1007/978-3-319-38992-9

MapReduce: software for parallel processing huge amount of data.

Flume: service to gather, aggregate, and move chunks of data from several sources to a centralized system.

Cassandra: an open source database system for analyzing large amount of data on a distributed system. It is characterized by a high performance and by a high availability with no single point of failure (i.e., a part of system that if fails stop the whole system). It fosters data denormalization, which means grouping data or adding redundant information, in order to optimize the database performance.

Distributed System: Multiple terminals communicating between them. The problem is divided in many tasks, and assigned to each terminal. It is a highly scalable system as further nodes are added.

Google File System: proprietary distributed file system for managing efficiently large datasets.

HBase: an open source non-relational database (column-oriented) developed on a HDFS. It is very useful for real time random read and write access to data, as well as to store sparse data (small specific chunk of data within a vast amount of them). The relational counterpart is called **Big Table**.

Enterprise Data Warehouse (EDW): system used for analysis and reporting that consists of central repositories of integrated data from a wide spectrum of different sources. The typical form of an EDW is the **extract-transform-load (ETL)**, that is the most representative case of *bulk data movement*, but other three important examples of these systems are **data marts** (i.e., a subset of the EDW extracted out in order to address a specific question), **Online analytical processing (OLAP)**—used for multidimensional low-frequency analytical query—and **Online transaction processing (OLTP)**—used rather for high volume fast transactional data processing. The wider system that includes instead a set of servers, storage, operating systems, database, business intelligence, data mining, etc. is called **data warehouse appliance (DWA)**.

Resilient Distributed Datasets (RDD): logical collection of data partitioned across machines. The most known examples is **Spark**, an open source clustering computing that has been designed to accelerate analytics on Hadoop thanks to the multi-stage in-memory primitives (that are basic data types defined in programming languages or built it with their support). It seems to run 100 times faster than Hadoop, but its disadvantage is that it does not provide its own distributed storage system.

Hive: additional example of EDW infrastructure that facilitates data summarization, ad-hoc queries, and specific analysis.

Pig: platform for processing huge amount of data through a native programming language called Pig Latin. It runs at the same time sequences of MapReduce.

Programming language: is a formal constructed language designed to communicate instructions to a machine. The main ones for data science applications are Java, C, C++, C#, R, and Matlab. Scala is another language that is becoming extremely popular right now, but it is an example of *functional language*.

Scripting Language: is a programming language that supports scripts, which are piece of codes written for a run-time environment that interpret (rather than compile) and automate the execution of tasks. The main ones in big data field are Python, JavaScript, PHP, Perl, Rub, and Visual Basic Script.

Data Mart: is a subset of the data warehouse used for a specific purpose. Data marts are then department-specific or related to a single line of business (LoB). The next level of data marts is the **Virtual Data Marts**, i.e., a virtual layer that create various views of data slices—in other words, instead of physically creating a data mart, it just takes a snapshot of them. The final evolution is instead called **Data Lakes**, which are massive repositories of unstructured data with an incredible computational capability. Hence, data marts physically create repositories (slices) of data, virtual data marts leave the data where they are and create virtual constructs—reducing the cost of transferring and replicating them—while data lakes work as the virtual data marts but with any kind of data format.

Appendix II—Data Science Maturity Test

The following questionnaire provided could help managers to grasp a rough idea of the current data stage of maturity they are facing within their organizations. It has to be integrated with deep conversations and meetings with the big data analytics (BDA) staff, the IT team, and supported by solid researches.

(1) What is your investment level in BDA capabilities?

1. Absent. We don't have money for big data
2. A small budget is allocated when positive quarters in core activities allow us to do that
3. A modest funding scheme is in place
4. We invested a good percentage of our revenues in BDA in the last year, and we will keep investing because it is part of our company's vision

(2) What executives' support to analytics capabilities looks like?

1. Neither IT nor business think BDA is useful to the business
2. Only IT managers support it because they are interested in the technological challenge
3. Business managers see the hidden value in data and support BDA projects
4. Both IT and business executives believe in BDA potential

(3) What is your current stage of working with data?

1. We will start using data in the future if needed
2. We have a good idea of what business questions we could solve with data in my company
3. We take action using analytics
4. We are automating analytics the most we can, and we believe is a competitive factor that gives us benefits we are able to communicate frequently to top management and shareholders

(4) Your analytics team is:

1. Inexistent
2. Acquired from outside at the moment
3. We have some senior scientist that has been recruited, but we are now growing the team internally by training
4. An independent sustainable group and function within the company

(5) Your company culture is:

1. Intolerant—especially for failure concerning new analytics, methodologies, and technologies
2. Variegated—it is half-half made by old-style professionals and geeks
3. Collaborative—people are willing to work together and share.
4. Creative—innovation is valued and we are encouraged and monetarily compensated for our original shared contributions.

(6) How your data science team is connected to the company hierarchy?

1. We only have some analysts with small tasks, who deliver the outcomes to their direct managers on a weekly/monthly basis
2. The data team is leaded by a business head, and their contribution is continuously marginally positive
3. Our data scientists are tight to our data warehouse and data management teams, and they constantly interconnected with the business side
4. They are autonomous and do not seat in the same building of the operations function. They are allocated in a Centre of Excellence.

(7) The internal data policy is:

1. Fairly poor, we do not need it
2. Metadata definitions and BDA policy are well-established
3. We have a BDA policy that we constantly monitor and we have a security policy for any data forms
4. We have a BDA and security policies, and we anonymized all the relevant data to protect our clients and partners' privacy.

(8) The data in your company are:

1. Stored in silos
2. We prioritized the data to be used within our organization, and they are internally shared
3. Many different data sources are integrated for our analysis, and we take care of data quality through a meticulous goodness assessment based either on the final use or the type of data we will exploit
4. We have integrated BDA technologies into our systems, we store our data on the cloud, and we often use them for mobile applications

Table A.1 Data science maturity test classification		Primitive	Bespoke	Factory	Scientific
	Score	10–15	16–25	26–35	36–40

(9) When your company looks at your BDA capabilities:

1. It sees mainly a sunk-cost, i.e., the cost of storing, maintaining, protecting and analyzing these datasets
2. We know data have value and we understand both the data cost and data competitive advantage, but we are definitely overwhelmed
3. We are rationalizing our data storage and usage abilities, because we understood that not everything is either pertinent or meaningful
4. We have an efficient process for data aggregation, integration, normalization and analysis, and we can manage easily any amount of inflowing data.

(10) Your firm is currently using:

1. Relational Database and Internal data
2. Data marts, R or Python languages, and public data
3. NoSQL database, Hadoop and MapReduce, and we use external data, sometimes also unstructured
4. Highly unstructured data, APIs, and a Resilient Database

Once each single question has been answered, it is simple to obtain a rough measure of the data maturity stage for a certain company. For each answer indeed, it has to be considered the number associated to that answer, and then it is enough to sum up all the numbers obtained in this way. So, for example, if in the third question the answer is "we take action using analytics", the number to be considered is 3, since it is the third answer of the list.

Finally, the score obtained should range between 1 and 40. The company will then belong to one of the four stages explained in Table 2.1 accordingly to the score achieved, that is explained in the table (Table A.1).

Appendix III—Data scientist Extended Skills List

Programming: R, Python, Scala, JavaScript, Java, Ruby, C++, C#

Statistics and Econometrics: probability theory, ANOVA, MLE, regressions, time series, spatial statistics, Bayesian Statistics (MCMC, Gibbs sampling, MH Algorithm, Hidden Markov Model), Simulations (Monte Carlo, agent-based modeling, NetLogo)

Scientific approach: experimental design, A/B testing, technical writing skills, RCT

Machine Learning: supervised and unsupervised learning, CART, algorithms (Support vector Machine, PCA, GMM, K-means, Deep Learning, Neural Networks)

Mathematics: Matrix algebra, relational algebra, calculus, optimization (linear, integer, convex, global)

Big Data Platforms: Hadoop, Map/Reduce, Hive, Pig, Spark, Storm

Text mining: Natural Language Processing, LDA, LSA, Part-of-speech tagging, Parsing, Machine Translation

Visualization: graph analysis, social networks analysis, Tableau, ggplot, D3, Gephi, Neo4j, Alteryx

Business: business and product development, budgeting and funding, project management, marketing surveys, domain/sector knowledge

Systems Architecture and Administration: DBA, SAN, cloud, Apache, RDBMS

Dataset Management:

- **Structured Dataset:** SQL, JSON, BigTable
- **Unstructured Dataset:** text, audio, video, BSON, noSQL, MongoDB, CouchDB
- **Multi-structured Dataset:** IoT, M2M

Data Analysis: feature extraction, stratified sampling, data integration, normalization, web scraping, pattern recognition

Appendix IV—Data Scientist Personality Questionnaire

The terminology used to classify into 16 subcategories the different kind of data scientists is given by the two-entry matrix exhibited in the Table 4.1. The terminology can be sometime misleading if related to the Keirsey Temperament Sorter (KTS), and this is why it is necessary to specify that the only categorization borrowed from KTS framework is the broader one, i.e. the Artisan-Idealist-Rational-Guardian partition. Every sub-category has instead to be taken as newly generated.

Here it follows the personality test to sort data scientists into a specific box. It is composed by 10 questions, and for each one a single answer has to be provided. This test is not a professional temperament test to fully understand individuals' personality, but it is more a quick tool for managers to efficiently and consciously allocate the right people to the right team.

(1) When you start working on a new dataset

    a. You start exploring immediately and querying the data
    b. Plan in advance how to tackle it

   c. You spent time in understanding the data, where they come from, and their meaning

   d. You identify a research question quickly, and focus on designing the a new improved method for analyzing your data

(2) In your team, people count on you for your

   a. Troubleshooting ability

   b. Organizational skills

   c. Capacity to reduce the problem complexity

   d. Strategic approach and conceptualization of the problem

(3) When facing a new data challenge, your first thought is

   a. Is what I am doing impactful and relevant?

   b. When do I have to deliver some results?

   c. How this challenge can make me better?

   d. What I can learn from this dataset?

(4) In a data analysis, which is the most important thing to you

   a. Results, no matter how you do achieve them, what strategy or technology you do employ

   b. To achieve a result in the correct way and with the right process or technology

   c. Attaining significant results in an ethical manner

   d. Reaching the outcomes through an accurate, replicable, and efficient procedure

(5) If you have finished your assigned today's daily, you would

   a. Focus again on your analysis and try to find alternative and innovative way to achieve your final goal

   b. Start with something else, even if this might mean to stay longer at your desk

   c. Help a colleague in difficulty with his analysis

   d. Give suggestions and highlight weaknesses in your colleagues' works for the sake of the team and business development

(6) If you would have some spare time during your daily work, you would prefer to

   e. Optimize existing technology for the whole company

   f. Improve your analysis

   g. Try to derive new insights from your previous analysis

   h. Understanding how to maximize the value of your analysis

(7) It is your data-dream of

     e. Speaking about data with only engineers and IT team
     f. Teaching data related contents
     g. Engaging with people who do not know anything about data science
     h. Persuading and convincing the business team of the big data opportunity

(8) You prefer to work with

     e. Huge amount of structured data
     f. Any kind of data that challenge me
     g. Behavioral or social media data, or any unusual data
     h. No data in particular

(9) If you would quit tomorrow your data science job, you would prefer to become

     e. An IT manager or software engineer
     f. A professor
     g. A consultant
     h. An entrepreneur

(10) What characteristic of big data you value the most

     e. Volume
     f. Velocity
     g. Variety
     h. Value

Once each question has been answered with a single choice, the result is given by pairing the reply chosen more often within the first five questions (a–d) with the answer that appears more often in the last five (e–f), as shown in the following table. So, if for instance in the first five questions *b* emerges as predominant answer, while in the last five *f* is the median, the person considered is a *Cruncher* (Table A.2).

Table A.2 Data scientist personality classification

Archetype/ personality	Artisan	Guardian	Idealist	Rational
Technical	*Gardener*: A–E	*Architect*: B–E	*Evangelist*: C–E	*Wrangler*: D–E
Researcher	*Alchemist*: A–F	*Cruncher*: B–F	*Champion*: C–F	*Groundbreaker*: D–F
Creative	*Trailblazer*: A–G	*Catalyst*: B–G	*Visionary*: C–G	*Warlock*: D–G
Strategist	*Babelian*: A—H	*Mastermind*: B–H	*Advocate*: C–H	*Fisherman*: D–H

Appendix V—Code of Professional Conduct—Instructions

1.   Terminology

Terms as data, data scientist, big data, have to be defined.

2.   Working Relationship

This section highlights the relevance of defining the scope of the relationship, and the means through which the scope has to be reached.

It has to guarantee as well professionalism, competences, independence and objectivity (Boyd and Crawford 2012).

A subparagraph has to explain how to proceed in case of misrepresentation, misconduct, or fraud.

Finally, a section on the quality of the analysis and results have to be presented, as well as how a data scientist should act in case the results he achieved are afterwards misrepresented or misused. He has to use diligence, scientific method, replicability of process and analysis, and not provide evidences he knows to be false or incomplete.

3.   Conflict of Interests

It shall explain the policy regarding disclosure of conflicts and limitations. Exceptions have to be listed.

4.   Duties to Clients

The data scientist has to present correctly his results, preserve the confidentiality of the agreement, and act for the benefit of his clients before his own or his employer's one.

It should include a section related to the communication with clients, as well as the disclosure of risks on relying on the (data) results obtained. The results should also be evaluated with reasonable diligence and explained to the extent of allowing the client to reach a decision by his own.

It would apply to current and prospective clients, with a series of further confidentiality in the second case (not revealing information from a prospective client, measures to avoid conflict of interests, etc.).

5.   Duties to Employers

In this section it has to be deal with themes as loyalty to the employer, and supervising responsibilities.

6   Confidential Information

This paragraph should define confidential information, setting the guidelines for protecting them, when the confidentiality can be breached (fraud, in order to prevent death, etc.), as well as the final return at the end of the project.

References

Boyd, D., & Crawford, K. (2012). Critical questions for big data: provocations for a cultural, technological, and scholarly phenomenon. *Information, Communication & Society, 15*(5), 662–679.

Dull, T. (2014). *A Non-Geek's Big Data Playbook*. SAS Best Practices White paper. Retrieved from http://www.sas.com/content/dam/SAS/en_us/doc/whitepaper1/non-geeks-big-data-playbook-106947.pdf.

Printed in the United States
By Bookmasters